DRAGON BLOOD TREE

ASHLEY GISH

look!

DRAGON BLOOD TREE

CREATIVE EDUCATION • CREATIVE PAPERBACKS

Published by Creative Education
and Creative Paperbacks
P.O. Box 227, Mankato, Minnesota 56002
Creative Education and Creative Paperbacks are imprints of
The Creative Company
www.thecreativecompany.us

Design and art direction by Blue Design
Edited by Kremena Spengler

Photographs by Getty Images/John M Lund Photography Inc, 1, 17, Srinophan69, 8, Teuku Boyhaqie, 11, 16; Unsplash/Andrew Svk, 9, 14–15, 18–19, 20; Wikimedia Commons/Cockburn del., 12, Fir0002, 15, John Nugent Fitch, 14, Krzysztof Ziarnek, Kenraiz, 10, Maša Sinreih in Valentina Vivod, 23, Rod Waddington, cover, 2, 5, 6–7, 13, 21, 22, 23

Library of Congress Cataloging-in-Publication Data
Names: Gish, Ashley, author.
Title: Dragon blood tree / Ashley Gish.
Description: Mankato, Minnesota : Creative Education and Creative Paperbacks, [2026] | Series: Look! | Includes bibliographical references and index. | Audience: Ages 6–9 | Audience: Grades 2–3 | Summary: "An exotic tree shaped like an umbrella, with blood-red sap? Elementary-level readers will learn all about the tropical dragon blood tree with this striking introduction to the unusual plant's life cycle"– Provided by publisher.
Identifiers: LCCN 2024044289 (print) | LCCN 2024044290 (ebook) | ISBN 9798889895862 (library binding) | ISBN 9781682777527 (paperback) | ISBN 9798889896661 (ebook)
Subjects: LCSH: Dracaena cinnabari–Juvenile literature. | Dracaena cinnabari–Life cycles–Juvenile literature.
Classification: LCC QK495.A83 G57 2026 (print) | LCC QK495.A83 (ebook) | DDC 584/.79–dc23/eng/20250109
LC record available at https://lccn.loc.gov/2024044289
LC ebook record available at https://lccn.loc.gov/2024044290

Printed in India

TABLE OF CONTENTS

A strange tree grows in just one place on Earth. It's shaped like an upside-down umbrella. Its **sap** is red like blood.

Dragon blood trees grow on Socotra Island in the Indian Ocean. They can live for more than 600 years. Their sap has been used to make dye and medicine.

sap: liquid that carries food and water through a plant

The dragon blood tree is named for its red sap.

10

The seeds begin to sprout after four to ten weeks.

Dragon blood trees grow from seeds. First, a root grows down into the soil. Then a stem pushes up above the ground. The first leaf grows from the stem.

The first leaves are pale and limp. Then they get more light. They turn green. The **sapling's** trunk is thick. It's round like a ball.

sapling: a baby tree

Saplings may grow just an inch (2.5 centimeters) in five years. It takes 100 years for saplings to grow to 4 feet (1.2 meters). Fully grown trees can be up to 50 feet (15.2 m) tall.

As the saplings grow into big trees, animals visit. Geckos, chameleons, and snakes make a home in the trees. Birds build nests among the branches.

Dragon blood trees are evergreens. They never lose their leaves.

Dragon blood trees grow flowers. But they may be 300 years old before this happens! The small flowers are white or green. They grow in bunches among the leaves.

Pollinators land on the flowers. They carry yellow powder called pollen from one flower to another. Because of the pollen, the flowers will grow fruit with seeds inside.

pollinator: insect that spreads pollen from one flower to another

The flowers falls off, and the fruit grows. The fruit starts out green. Then it turns black. The ripe fruit is red or orange. Each fruit has up to four seeds inside.

Birds eat the fruit. But they poop out the seeds. This is how seeds are spread across Socotra.

DRAGON BLOOD TREE FRUIT

Socotra starlings eat the fruit of dragon blood trees.

Socotra is an **arid** island. But the dragon blood tree's shape allows water to flow to the middle of the tree. It keeps the water from quickly drying up around the trunk.

Human activity is the main threat to dragon blood trees. People cut them down for their wood and sap. Goats and other farm animals eat the saplings.

arid: dry and with little to no rain

Until people brought farm animals, bats were the only mammals on Socotra.

20

Dragon blood trees make homes for all kinds of animals.

The dragon blood tree starts as a seed. It gets planted in the ground. Its roots grow down. The first leaf grows up. Hundreds of years pass. The tree makes flowers. Pollinators visit the flowers. Fruit grows. Birds eat it and spread the seeds. The seeds sprout. If animals do not eat the saplings, they will grow into trees. The trees will live for 600 years or more!

Life Stages

Seed - Root - First leaf - Sapling - Tree - Flower - Fruit

Dragon blood trees have red sap. If the tree is hurt, the sap hardens into **resin**. Resin smells good when it is burned. It is used to add color to makeup and musical instruments. People have collected the resin for hundreds of years. The only way to get resin is to cut into the tree. This may kill the tree.

resin: *sticky material made by trees*

WEBSITES

Dragon Blood Tree
https://easyscienceforkids.com/dragon-blood-tree
Learn more facts about the dragon blood tree.

Top Ten Oddest Looking Plants
https://www.earthrangers.com/EN/CA/top-10/top-ten-oddest-looking-plants
Learn about the dragon blood tree and other odd plants.

READ MORE

Kaiser, Brianna. *Weird Plants.* Minneapolis: Lerner Publishing, 2023.

Levy, Janey. *Dragon's Blood Trees Bleed!* New York: Gareth Stevens Publishing, 2019.

Lundgren, Julie K. *Creepy but Cool Scary Plants.* New York: Crabtree Seedlings, 2021.

INDEX